Abbou Ahmed

Steady state Analysis, DC voltage control of SEIG used in wind turbine

AF153285

Abbou Ahmed

Steady state Analysis, DC voltage control of SEIG used in wind turbine

LAP LAMBERT Academic Publishing

Impressum / Imprint
Bibliografische Information der Deutschen Nationalbibliothek: Die Deutsche Nationalbibliothek verzeichnet diese Publikation in der Deutschen Nationalbibliografie; detaillierte bibliografische Daten sind im Internet über http://dnb.d-nb.de abrufbar.
Alle in diesem Buch genannten Marken und Produktnamen unterliegen warenzeichen-, marken- oder patentrechtlichem Schutz bzw. sind Warenzeichen oder eingetragene Warenzeichen der jeweiligen Inhaber. Die Wiedergabe von Marken, Produktnamen, Gebrauchsnamen, Handelsnamen, Warenbezeichnungen u.s.w. in diesem Werk berechtigt auch ohne besondere Kennzeichnung nicht zu der Annahme, dass solche Namen im Sinne der Warenzeichen- und Markenschutzgesetzgebung als frei zu betrachten wären und daher von jedermann benutzt werden dürften.

Bibliographic information published by the Deutsche Nationalbibliothek: The Deutsche Nationalbibliothek lists this publication in the Deutsche Nationalbibliografie; detailed bibliographic data are available in the Internet at http://dnb.d-nb.de.
Any brand names and product names mentioned in this book are subject to trademark, brand or patent protection and are trademarks or registered trademarks of their respective holders. The use of brand names, product names, common names, trade names, product descriptions etc. even without a particular marking in this work is in no way to be construed to mean that such names may be regarded as unrestricted in respect of trademark and brand protection legislation and could thus be used by anyone.

Coverbild / Cover image: www.ingimage.com

Verlag / Publisher:
LAP LAMBERT Academic Publishing
ist ein Imprint der / is a trademark of
OmniScriptum GmbH & Co. KG
Heinrich-Böcking-Str. 6-8, 66121 Saarbrücken, Deutschland / Germany
Email: info@lap-publishing.com

Herstellung: siehe letzte Seite /
Printed at: see last page
ISBN: 978-3-659-80955-2

Copyright © 2015 OmniScriptum GmbH & Co. KG
Alle Rechte vorbehalten. / All rights reserved. Saarbrücken 2015

Steady state, Analysis and DC voltage control of Self excited induction generator used in wind turbine

Ahmed ABBOU

Ahmed ABBOU is with the Mohammadia School of Engineers, Mohammed V University Rabat Morocco; (e-mail: abbou@ emi.ac.ma).

AHMED ABBOU

Steady state, Analysis and DC voltage control of Self Excited Induction Generator used in wind turbine

Effects of variation of excitation capacitance

Control strategy to maintain the terminal voltage constant

2

TABLES OF CONTENTS

NOMENCLATURE

$V_{s\alpha,\beta}$: Stator voltage in the stationary α, β axis

V_{dc} : DC output voltage

i_{dc} : Total DC current

$i_{s\alpha,\beta}$: Stator current in the stationary α, β axis

$\Phi_{r\alpha,\beta}$: Rotor flux in the stationary α, β axis

Ω : Rotor speed

ω_r : Rotor electrical speed

σ : Leakage coefficient of the machine

C_r : Load torque

J : Rotor inertia

p : Number of pole pairs

R_s : Stator resistance

R_r : Rotor resistance

R_{dc} : DC load resistance

C_{dc} : Output capacitor

L_{sg} : Stator inductance of induction generator

L'_{rg} : Rotor inductance of induction generator

L_{mg} : Mutual inductance of induction generator

K_d, K_q: are constants which represent the initial induced voltages along the d-axis and q-axis respectively due to remanent magnetic flux in the core.

M : Mutual inductances of induction motor

L_s : Stator cyclic inductance of induction motor

L_r : Rotor cyclic inductance of induction motor

Tr : Rotor time constant

Ts : Stator time constant

Sa, Sb, Sc : Switching functions of the three inverter legs

s : Laplace operator

4

LISTES OF FIGURES

LISTES OF TABLES

7

INTRODUCTION

Recently considerable attention is being focused on environmentally clean and safe renewable energy sources like wind, solar, hydro etc. Many types of generator concepts have been used and proposed to convert wind power into electricity. The size of the wind turbines has increased during the past ten years, and the cost of energy generated by wind turbine has decreased. The challenge is to build larger wind turbines and to produce cheaper electricity.

In a standalone induction generator the major problem is that of guaranteeing self-excitation. Self-excitation of an induction machine and its sustenance depend on the appropriate combination of speed, load and terminal capacitance in relation to the magnetic non-linearity of the machine. These in turn cause certain limitations on the performance of the machine. In view of these, studies on the criteria for self-excitation of an induction generator are considered to have practical significance. The excitation requirements, of an induction generator have been dealt with extensively in the literature [1].

For self-excitation to occur, the following two conditions must be satisfied:

- The rotor should have sufficient residual magnetism.
- The three capacitor bank should be of sufficient value.

If an appropriate capacitor bank is connected across the terminals of an externally driven induction machine and if the rotor has sufficient residual magnetism an EMF is induced in the machine windings due to the excitation provided by the capacitor. The EMF if sufficient would circulate leading currents in the capacitors. The flux produced due to these currents would assist the residual magnetism. This would increase the machine flux and larger EMF will be induced. This in turn increases the currents and the flux. The induced voltage and the current will continue to rise until the VAR supplied by the capacitor is balanced by the VAR demanded by the machine, a condition which is essentially decided by the saturation of the magnetic circuit. This process is thus cumulative and the induced voltage keeps on rising until saturation is reached. To start with transient analysis the dynamic modeling of induction motor has been used which further converted into induction generator [2]-[4]. Magnetizing inductance is the main factor for voltage buildup and stabilization of generated voltage for unloaded and loaded conditions. The dynamic Model of Self Excited Induction Generator is helpful to analyze all characteristic especially dynamic characteristics. For the past few years the researches has been developed positively in the steady state models of three phase self-excited induction generator(SEIG) [5] and proposed the steady state equivalent circuit which represents the SEIG, the critical capacitance requirement and excitation balancing has been proposed [5]-[6]. Accordingly the better applicability of induction motor as a generator for isolated applications has been proposed [7]. The model was found suitable for steady state analysis but not transient analysis. Thus for analyzing the transient characteristics, dynamic model of SEIG has been developed [8] and analyzed the dynamic characteristics for various transient conditions and stability.

To regulate the voltage of a SEIG with changing load and speed, self-excitation capacitors may be supplemented with an active external source of reactive power. The function of the voltage regulator is to maintain the output voltage of the generator within a given operating range.
Different methods have been proposed in the literature for regulating the voltage of the SEIGs [12]. The scheme based on switched capacitors [13] finds limited application because it regulates the terminal voltage in discrete steps and it may create switching transients. In [14], the author have used a hybrid excitation unit consisting of a capacitor bank and an active power filter to regulate the output voltage of stand-alone SEIG and proposed the advanced deadbeat current control strategy that works with variable speed to reduce the system cost.

On other hand, Fuzzy control was found particularly useful to solve non-linear control problems or when the plant model is unknown or difficult to build [15]. In this work it will be shown that these techniques can also be useful in applications where classical control performs well. Fuzzy Logic allows a simpler and more robust control solution whose performance can only be matched by a classical controller with adaptive characteristics, much more difficult to implement.

In this context, this work presents generalized state-space dynamic model of a three phase SEIG developed using d-q variables in stationary reference frame for transient analysis. The proposed model for induction generator, load and excitation using state space approach driving by wind turbine and supplying which is coupled to a centrifugal pump in order to optimize his performances.

Also we purpose the fuzzy DC voltage control with direct torque and flux control strategy of an isolated self-excited induction generator (SEIG) for DC power application. The single DC side capacitor provides all the reactive current or the VAR required by the generator and the load. The controller is based on Fuzzy Logic (FL) which provides high dynamics performances both in transient and steady state response.

Chapter I

Steady state and analysis of self-exited induction generator

I. Steady state and analysis of self-exited induction generator

I.1 Systems Modeling

I.1.1 Proposed system

In the proposed system (Figure 1), a power generation system consisting of a wind turbine with Self-excited induction generator (SEIG).

The produced power is used to supply an induction motor coupled to a centrifugal pump. As the SEIG requires reactive power for its excitation, a three phase capacitor bank is connected across its stator terminals.

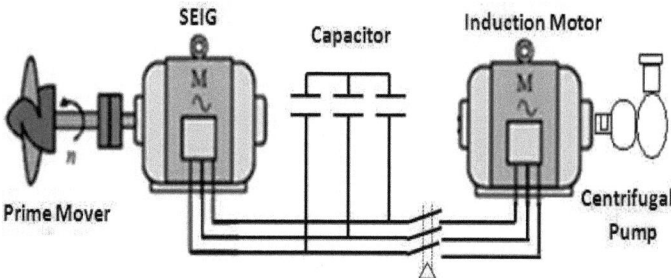

Figure 1: Wind electric pumping system

The Induction Motor cannot be supplied unless the SEIG stator voltage build up process occurs. For this reason an operating mode switcher selects first the no load condition until the voltage build up process is accomplished. Subsequently the switcher is turned on as to connect the Induction Motor to the SEIG.

In order to analyze the performances of self-excited induction generator which supplies an induction motor driving pump, a system modeling is required. Following, a steady state and dynamic modeling are presented.

I.1.2 Self-Exited Induction Generator Modeling

The model for the SEIG is similar to that of the induction motor. To model the SEIG effectively, the parameters should be measured accurately. The parameters used in the SEIG can be obtained by conducting tests on the induction generator when it is used as a motor. The traditional tests used to determine the parameters are the open circuit (no load) test and the short circuit (locked rotor) test.

In this work the d-q model is used because it is easier to get the complete solution, transient and steady state, of the self-excitation. The parameters given in the d-q equivalent circuit shown in Figure 2 are obtained by conducting parameter determination tests on the above mentioned induction machine. As it is a wound rotor induction machine there is no variation of rotor parameters with speed.

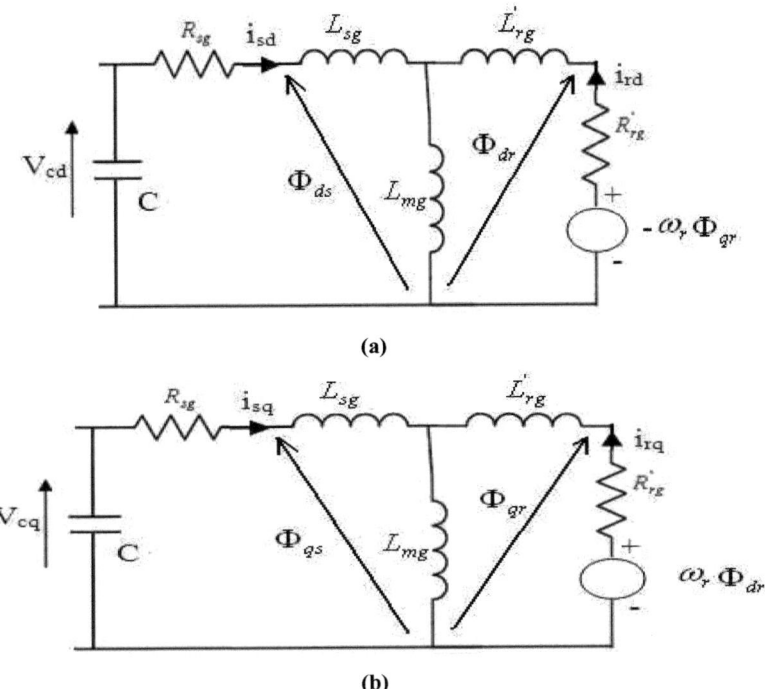

(a)

(b)

Figure 2: d-q model of SEIG at no load a) d-axis, b) q-axis

The parameters obtained from the test at rated values of voltage and frequency are $L_{sg}=L'_{rg}=229mH$, $L_{mg}=217mH$, $R_{sg}=2.2\Omega$, $R'_{rg}=2.68\Omega$. For motoring application these parameters can be used directly.

However, for SEIG application the variation of L_{mg} with voltage should be taken into consideration.

$$
\begin{bmatrix}
R_{sg}+pL_{sg}+\dfrac{1}{pC} & 0 & pL_{mg} & 0 \\[2mm]
0 & R_{sg}+pL_{sg}+\dfrac{1}{pC} & 0 & pL_{mg} \\[2mm]
pL_{mg} & -\omega_r L_{mg} & R_{rg}+pL_{rg} & -\omega_r L_{rg}^{'} \\[2mm]
\omega_r L_{mg} & pL_{mg} & \omega_r L_{rg}^{'} & R_{rg}+pL_{rg}
\end{bmatrix}
\begin{bmatrix}
i_{sq} \\ i_{sd} \\ i_{rq} \\ i_{rd}
\end{bmatrix}
=
\begin{bmatrix}
0 \\ 0 \\ 0 \\ 0
\end{bmatrix}
\tag{1}
$$

The initial conditions for self-excitation, namely the remanent magnetic flux in the rotor and/or the initial charge in the capacitors are not considered because they will be cancelled when both sides are differentiated.

Derived from Equation (1) and including initial conditions, i.e. initial voltage in the capacitors and remanent magnetic flux in the core, one can obtain the following differential equation [5]:

$$
pI = AI + B \tag{2}
$$

Where:

$$
I =
\begin{bmatrix}
i_{sq} \\ i_{sd} \\ i_{rq} \\ i_{rd}
\end{bmatrix}
\qquad
B = \frac{1}{L}
\begin{bmatrix}
L_{mg}K_q - L_{rg}^{'}V_{cq} \\
L_{mg}K_d - L_{rg}^{'}V_{cd} \\
L_{mg}V_{cq} - L_{sg}K_q \\
L_{mg}V_{cd} - L_{sg}K_d
\end{bmatrix}
$$

$$
A = \frac{1}{L}
\begin{bmatrix}
-L_{rg}^{'}R_{sg} & -L_{mg}^2\omega_r & L_{mg}R_{rg} & -L_{mg}L_{rg}^{'}\omega_r \\
L_{mg}^2\omega_r & -L_{sg}R_{sg} & L_{mg}L_{rg}^{'}\omega_r & L_{mg}R_{rg} \\
L_{mg}R_{sg} & L_{sg}L_{mg}\omega_r & -L_{sg}R_{rg} & L_{sg}L_{rg}^{'}\omega_r \\
-L_{sg}L_{mg}\omega_r & L_{mg}R_{sg} & -L_{sg}L_{rg}^{'}\omega_r & -L_{sg}R_{rg}
\end{bmatrix}
$$

And $\quad L = L_{sg}L_{rg}^{'} - L_{mg}^2$

K_d and K_q are constants which represent the initial induced voltages along the d-axis and q-axis respectively due to remanent magnetic flux in the core.

To approach the characteristics of the induction machine (All the experimental points L_{mg}) by a mathematical function, we used an approximation method.

I.1.3 Characteristics of magnetising inductance in induction machine

In the modelling of an induction machine used for motoring applications, it is important to determine the magnetising inductance, L_{mg}, at rated voltage. In the SEIG the variation of magnetising inductance is the main factor in the dynamics of voltage build up and stabilisation. In this investigation the magnetising inductance is determined by driving the induction machine at synchronous speed and taking measurements when the applied voltage was varied from zero to 120% of the rated voltage with rated frequency. The magnetising inductance is calculated, without approximation, using the parameter presented in Table 4 in appendix. Here, conventional high accuracy meters are used for measurements of voltage, current and power, because the accuracy of the voltage and current sensors in the fast measurement system are not good for low values (close to zero) of voltages and currents. The computed power will be erroneous if the accuracy of voltage and current measurements is poor. This is especially important because the magnetising inductance for voltages and currents close to zero is used in the calculation for the initiation of self-excitation process.

The variation of the magnetising inductance, measured at rated frequency, for the induction machine used in this investigation is given in Figure 3.

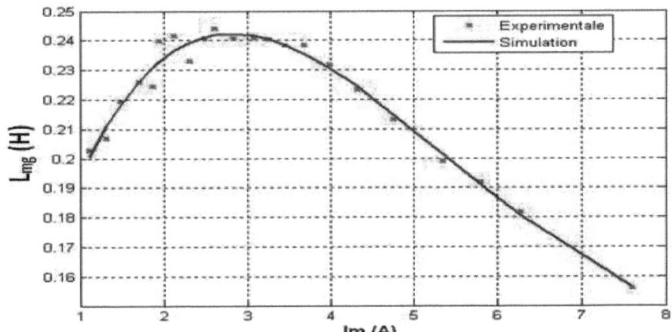

Figure 3: magnetizing inductance of the induction generator

I.1.4 Induction motor model

The linear model of the induction machine is widely known and used. It yields results relatively accurate when the operating point studied is not so far from the conditions of the model parameter identification. The traditional tests used to determine the parameters are the open circuit (no load) test and the short circuit (locked rotor) test.

The state space model of induction generator in the synchronously rotating reference frame α and β axes are [16]:

$$\begin{cases} \dot{X} = A.X + B.U \\ Y = C.X \end{cases} \tag{3}$$

Where A, B and C are the evolution, the control and the observation matrices respectively.

$$X = [i_{s\alpha} \quad i_{s\beta} \quad \Phi_{s\alpha} \quad \Phi_{s\beta}]$$

$$U = \begin{bmatrix} V_{s\alpha} \\ V_{s\beta} \end{bmatrix} \qquad ; \qquad Y = \begin{bmatrix} i_{s\alpha} \\ i_{s\beta} \end{bmatrix}$$

$$A = \begin{bmatrix} -\left(\dfrac{1}{\sigma T_s} + \dfrac{1-\sigma}{\sigma T_r}\right) & 0 & \dfrac{1-\sigma}{\sigma M T_r} & \dfrac{1-\sigma}{\sigma M}\omega \\ 0 & -\left(\dfrac{1}{\sigma T_s} + \dfrac{1-\sigma}{\sigma T_r}\right) & -\dfrac{1-\sigma}{\sigma M}\omega & \dfrac{1-\sigma}{\sigma M T_r}\omega \\ \dfrac{M}{T_r} & 0 & -\dfrac{1}{T_r} & -\omega \\ 0 & \dfrac{M}{T_r} & \omega & -\dfrac{1}{T_r} \end{bmatrix}$$

$$B = \begin{bmatrix} \dfrac{1}{\sigma L_s} & 0 \\ 0 & \dfrac{1}{\sigma L_s} \\ 0 & 0 \\ 0 & 0 \end{bmatrix} \qquad C = \begin{bmatrix} 1 & 0 & 0 & 0 \\ 0 & 1 & 0 & 0 \end{bmatrix}$$

With, ω Rotor speed and the machine's parameters: R_s, R_r are respectively the stator and the rotor resistance, M, L_s, L_r are respectively the mutual, the stator and the rotor cyclic inductances; p denotes the number of pole pairs, with:

$$T_r = \dfrac{L_r}{R_r}, \quad T_s = \dfrac{L_s}{R_s}, \quad \sigma = 1 - \dfrac{M^2}{L_s L_r}$$

The classical equations of the Induction Motor in the Park model are written follows:

$$\begin{cases} V_{ds} = R_{sm} i_{ds} + \dfrac{d\Phi_{ds}}{dt} - \omega_s \Phi_{qs} \\[2mm] V_{qs} = R_{sm} i_{qs} + \dfrac{d\Phi_{qs}}{dt} + \omega_s \Phi_{ds} \\[2mm] V_{dr} = 0 = R_{rm} i_{dr} + \dfrac{d\Phi_{dr}}{dt} - (\omega_s - \omega)\Phi_{qr} \\[2mm] V_{qr} = 0 = R_{rm} i_{qr} + \dfrac{d\Phi_{qr}}{dt} + (\omega_s - \omega)\Phi_{dr} \end{cases} \tag{4}$$

$$\begin{cases} \Phi_{ds} = L_{sm} i_{ds} + L_{mm} i_{dr} \\[1mm] \Phi_{qs} = L_{sm} i_{qs} + L_{mm} i_{qr} \\[1mm] \Phi_{dr} = L_{rm} i_{dr} + L_{mm} i_{ds} \\[1mm] \Phi_{qr} = L_{rm} i_{qr} + L_{mm} i_{qs} \end{cases} \tag{5}$$

V_{ds}, V_{qs}, V_{dr}, V_{qr}, i_{ds}, i_{qs}, i_{dr}, i_{qr} are respectively the voltage and current output of the Motor in the Park model.

R_{rm}, R_{sm}, L_{sm} and L_{rm} are respectively the resistance and inductances of the rotor and stator winding, $\omega = p\Omega_{mec}$ is the rotor speed, p is the number of pole pair.

The electromagnetic torque is given by the following formula:

$$T_{em} = p(\Phi_{ds} i_{qs} - \Phi_{qs} i_{ds}) \tag{6}$$

I.1.5 Pump model

The mechanical power of centrifugal pump is giving by:

$$P_{mec\_pump} = T_{pump} \Omega_{rm} = K.\Omega^2_{rm} \tag{7}$$

Where the T_{pump} is the load torque of the pump and K is a coefficient computed by:

$$K = \frac{T_{pumpMAX}}{\Omega_{rMAX}} \tag{8}$$

$T_{pumpMAX}$ is the maximum rated torque and Ω_{rMAX} is the maximum rated mechanical motor speed. The mechanical equation that describes this system is:

17

$$J_m \frac{d\Omega_{rm}}{dt} = T_{em} - T_{pump} \tag{9}$$

I.2 Steady state Analysis of Self Excited Induction Generator

I.2.1 Minimal capacitance

Figure 2 shows the per-phase equivalent circuit commonly used for SEIG supplying an induction motor. A three phase induction machine can be operated as a SEIG if its rotor is externally driven at a suitable speed and a three-phase capacitor bank of a sufficient value is connected across its stator terminals. When the induction machine is driven at the required speed, the residual magnetic flux in the rotor will induce a small electromotive force in the stator winding. The appropriate capacitor bank causes this induced voltage to continue to increase until an equilibrium state is attained due to magnetic saturation of the machine.

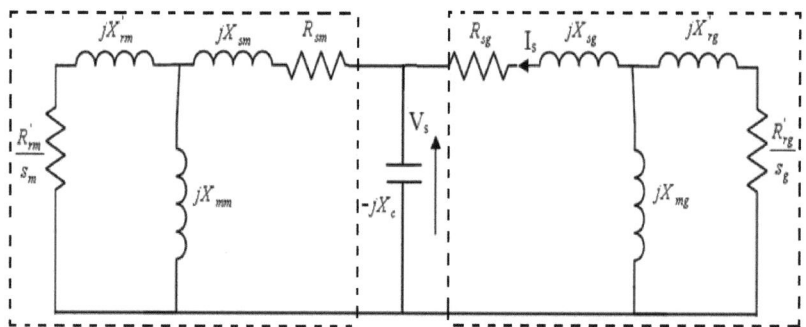

Figure 4: Per phase equivalent circuit of self-excited induction generator feeding an induction pump motor

We note: Index **g** for Induction Generator

Index **m** for Induction Motor

All circuit's parameters except the magnetizing inductance L_{mg} are assumed to be constant and insensitive to saturation.

From Figure 4, the total current at node a may be given by:

$$Vs.(Yg+Yc+Ym)=0 \qquad (10)$$

Where, Yg is a total admittance induction generator

   Yc is admittance capacitive

   Ym is a total admittance induction motor

If we denote: **a** P.U. frequency and **b** P.U. speed

So:

The expression of admittance capacitive is giving by:

$$Yc = j\frac{a}{Xc} \qquad (11)$$

The induction generator admittance is expressed as:

$$Yg = \frac{Yg_1(Yg_2 + Yg_3)}{Yg_1 + Yg_2 + Yg_3} \qquad (12)$$

With:

$$Yg_1 = \frac{1}{R_{sg} + jaX_{sg}}$$

$$Yg_2 = \frac{1}{jaX_{mg}} \qquad (13)$$

$$Yg_3 = \frac{1}{\dfrac{aR'_{rg}}{a-b} + jaX'_{rg}}$$

As a consequence of the symmetry of per phase equivalent circuit, the expression of total induction motor admittance Ym can be deduced from that of Yg by replacing the index **g** by **m**.

Therefore, under steady state self-excitation, the total admittance must be zero, since:

$$Vs \neq 0 \qquad \text{So} \qquad (Yg + Yc + Ym) = 0 \qquad (14)$$

Equation (14) is divided into real and imaginary parts as:

$$\Re(Yg + Yc + Ym) = 0 \tag{15}$$

$$\Im(Yg + Yc + Ym) = 0 \tag{16}$$

Separating real and imaginary Parts of the Yg, we obtain:

$$Yg = \frac{1}{R_G + jX_G} = \frac{R_G}{R_G^2 + (X_G)^2} - j\frac{X_G}{R_G^2 + (X_G)^2} \tag{17}$$

With:

$$R_G = R_{sg} + \frac{a(a-b)R'_{rg}X_{mg}^2}{(a-b)^2(X'_{rg} + X_{mg})^2 + R_{rg}'^2} \tag{18}$$

$$X_G = aX_{sg} + \frac{aX_{mg}((a-b)^2 X'_{rg}(X_{mg} + X'_{rg}) + R_{rg}'^2)}{(a-b)^2(X'_{rg} + X_{mg})^2 + R_{rg}'^2} \tag{19}$$

To simplify the equations and a nominal condition of the induction motor, we can write:

$$Ym = \frac{1}{R_M + jaX_M} = \frac{R_M}{R_M^2 + (aX_M)^2} - j\frac{aX_M}{R_M^2 + (aX_M)^2} \tag{20}$$

R_M and X_M are expressed with induction motor parameters.

Equation the real (15) and imaginary (16) parts to zero gives:

$$\frac{R_G}{R_G^2 + (X_G)^2} + \frac{R_M}{R_M^2 + (X_M)^2} = 0 \tag{21}$$

$$\frac{a}{Xc} - \frac{X_G}{R_G^2 + (X_G)^2} - \frac{X_M}{R_M^2 + (X_M)^2} = 0 \tag{22}$$

It is noted that (21) is independent of Xc and the only variable is the per unit frequency **a**. Once the value of a has been determined then Xc can be determined using (22).

For no load operation $R_M = \infty$ and $X_M = 0$

Substituting $R_M = \infty$ and $X_M = 0$ in (21):

$$R_{sg} + \frac{a(a-b)R_{rg}^{'}X_{mg}^{2}}{R_{rg}^{'}{}^{2} + (a-b)^{2}(X_{rg}^{'} + X_{mg})^{2}} = 0 \tag{23}$$

On simplification, it yields the following:

$$a_{max} = b - \frac{b}{2}\left[\frac{1 - \sqrt{1 - (\frac{b_c}{b})^{2}}}{1 + \frac{R_{sg}}{R_{rg}^{'}}(1 + \frac{X_{rg}^{i}}{X_{mg}})^{2}}\right] \tag{24}$$

Where b_c is given by:

$$b_c = \frac{2R_{sg}}{X_{ms}}\sqrt{\frac{R_{rg}^{'}}{R_{sg}} + (1 + \frac{X_{rg}^{'}}{X_{mg}})^{2}} \tag{25}$$

Substituting $R_M = \infty$ and $X_M = 0$ in (22):

$$Xc = a_{max}^{2}[X_{sg} + \frac{aX_{mg}((a-b)^{2}X_{rg}^{'}(X_{mg} + X_{rg}^{'}) + R_{rg}^{'}{}^{2})}{(a-b)^{2}(X_{rg}^{'} + X_{mg})^{2} + R_{rg}^{'}{}^{2}}] \tag{26}$$

Hence C_{min} is given by:

$$C_{min} = \frac{1}{2\pi 50.a_{max}^{2}(X_{sg} + \frac{aX_{mg}((a-b)^{2}X_{rg}^{'}(X_{mg} + X_{rg}^{'}) + R_{rg}^{'}{}^{2})}{(a-b)^{2}(X_{rg}^{'} + X_{mg})^{2} + R_{rg}^{'}{}^{2}})} \tag{27}$$

Thus, C_{min} is inversely proportional to the square of the p.u. machine frequency. The value of Cmin determined from (27) is just sufficient to have self-excitation under steady state. If a terminal capacitor $C = C_{min}$ is used and the generator is started from rest, the voltage build up will not take place.

Thus in practice, terminal capacitor C having a value somewhat greater than C_{min} should be selected to ensure self-excitation.

I.2.2 Dynamic results

The induction machine used as the SEIG in this investigation is a three-phase squirrel cage induction generator with specification: 3Kw, 220/380V, 12.4/7.2A, 50Hz.

This later supplies an induction motor: 1.5Kw, 220/380V, 8/5.6A, 50Hz.

The fixed parameters of both induction machines used in this proposed system are:

R_{sg}=2.2Ω, R'_{rg}=2.68Ω, L_{sg}=L'_{rg}=229mH, L_{mg}=f(I_m), p_g=2

R_{sm}=4.85Ω,R'_{rm}=3.80Ω,L_{sm}=L'_{rm}=274mH,L_{mm}=258mH, p_m=2.

The residual magnetism in the machine is taken into account in simulation process without which it is not possible for the generator to self-excite. Initial voltage in the capacitor is considered.

BASE SPEED=1500RPM

SPEED (PU)	FREQUENCY (PU)	SPEED (PU)	FREQUENCY (PU)
1	0.9987	0.6	0.5978
0.9	0.8986	0.5	0.4974
0.8	0.7984	0.4	0.3967
0.7	0.6982	0.3	0.2955

Table 1: Variation of frequency with speed

The critical speed b_c (25) is the speed below which the machine will not operate. For the given machine parameters R_{sg}, R'_{rg}, X_{sg}, X'_{rg}, speed b and magnetizing reactance X_{mg}, Equation (24) was solved to obtain the p.u. frequency a_{max} corresponding to self – excitation and the critical speed b_c was obtained from (25), for each value of p.u speed, the frequency a_{max} was be calculated. Table I. shows the variation of a_{max} for different p.u speeds b with initial value of the reactance X_{mg}=68.138Ω.

I.2.2.1 Effect of the capacitance value

In order to determine the suitable operation mode of SEIG and Induction motor system driven by wind turbine, the effect of the capacitance value on the stator voltage waveform is studied. The following steps resume a few applied tests to verify the system robustness:

First step: The rotor speed of SEIG is increased from zero at 0.1sec to 955 rpm at 0.5 sec.

2nd step: The motor pump starts operation after a delay of 4s.

Step 3: An additional load torque pump is applied at 6sec.

Step 4: The rotor speed of SEIG is decreased and attains 815rpm in 8sec.

The minimum value of the calculated capacitor under this test conditions is to 220µF.

Figures show that when the generator is excited with increasing values of capacitance, the steady state value of voltage and current generated has been increased. For small values of capacitance (C=240µF), there is a risk of losing the full excitation especially during transitional regimes (motor started, decrease rotor speed)

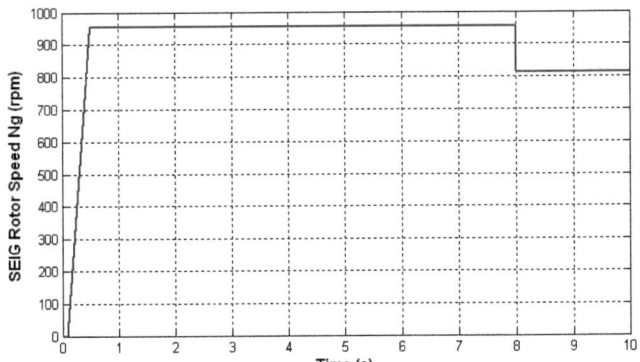

Figure 5: Evolution of SEIG rotor speed

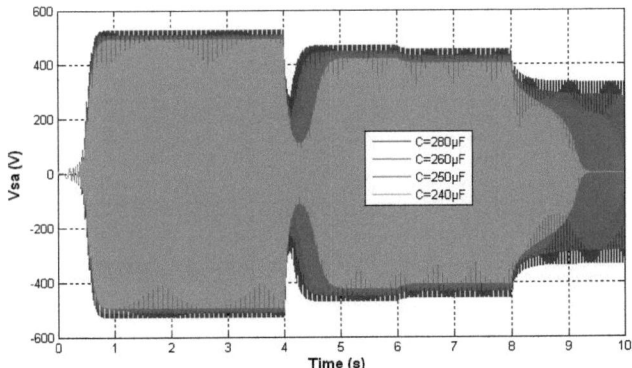

Figure 6: SEIG stator voltage variation at different capacitance

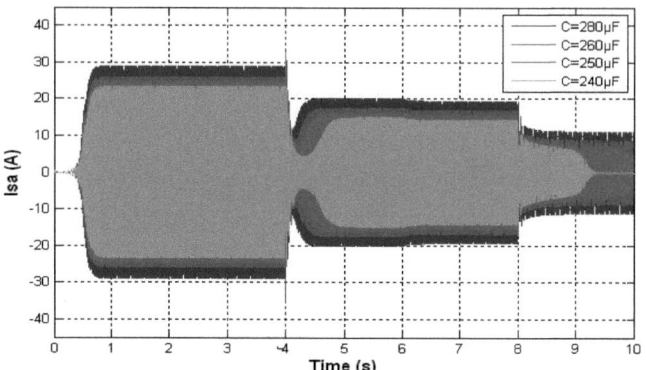

Figure 7: SEIG stator current variation at different capacitance

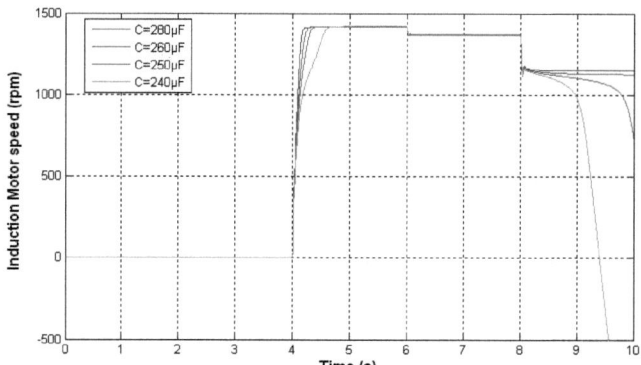

Figure 8: Induction motor speed variation at different capacitance

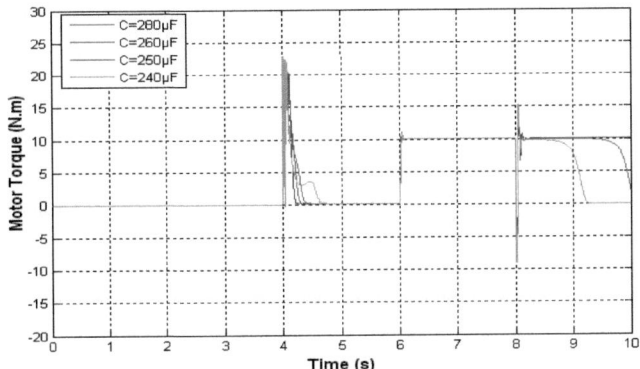

Figure 9: Induction motor Torque variation at different capacitance

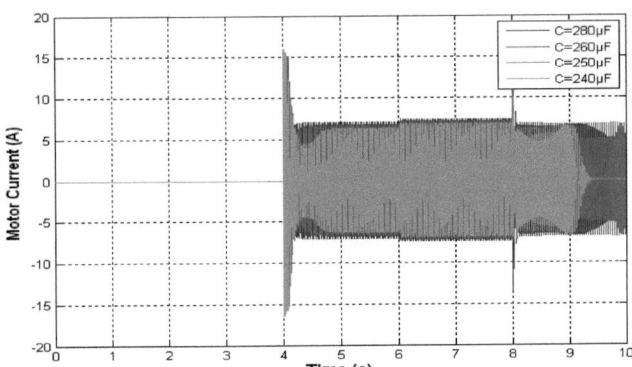

Figure 10: Induction motor current variation at different capacitance

From these figures, we see that the good performance of the motor pump are obtained with capacitance value equal to 260μF with maintaining the full excitation and suitable values of the voltage and current generated.

I.2.2.2 Effect of SEIG Rotor Speed variation with optimal capacitance

In order to verify the effect of speed rotor variation about functioning of the studied system, the generator is excited with capacitance value $C_{opt}=260\mu F$ and the rotor speed is increased from zero to Ng in 1sec. At t=4sec, the motor pump is started.

The following figures show that the voltage and current generated increases as the speed increases. The best speed that is suitable for rated motor pump is around at 1000 rpm.

With the slower machine it takes more time to full excitation. The full self excitation time is reduced from 1.7 sec to 0.9 sec, when the rotor speed increased from 763rpm to 1500rpm.

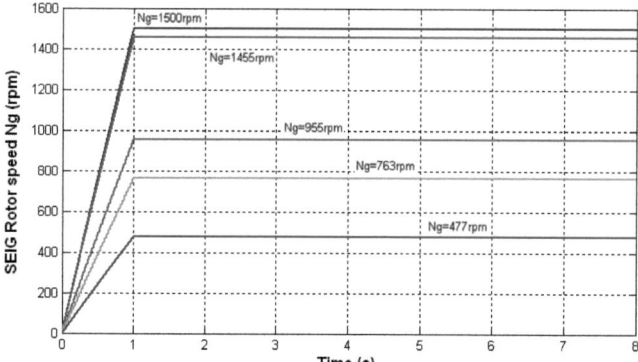

Figure 11: Evolution of SEIG rotor speed

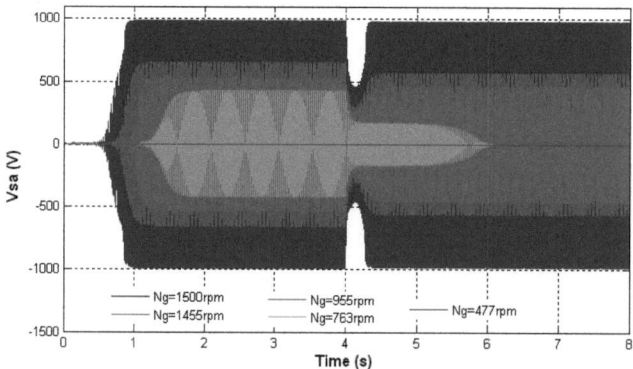

Figure 12: SEIG stator voltage variation at different rotor speed Ng

26

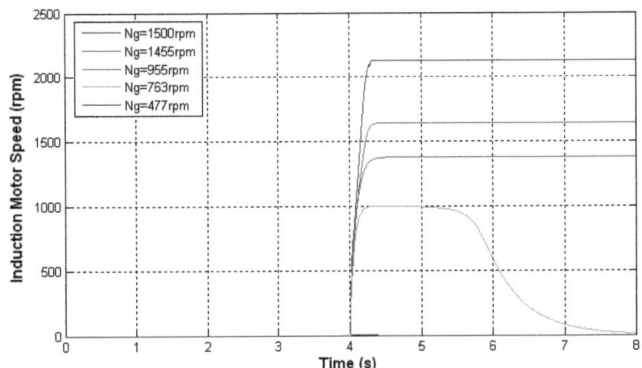

Figure 13: Induction motor speed variation at different rotor speed Ng

I.2.2.3 Summary

This chapter has presented generalized state space dynamic modeling of three phases self-excited induction generator. An effect of an excitation capacitor on the steady state behavior of the SEIG driven by wind turbine and supplying an induction motor which is coupled to a centrifugal pump is studied to enable selection of optimal capacitance value for a giving wind rotor. The approach presented enhances the transient characteristics with the variations of load torque, excitation and speed. Performance of SEIG coupled to induction motor system is analyzed during initial self-excitation, load torque switching, varying prime mover speed and excitation capacitance. It has been shown that a proper combination of speed and terminal capacitance can only guarantee self-excitation in induction generator and good functioning for motor pump. The experimental setup used to determine the magnetization characteristic can be used to verify the obtained curves and capacitances values calculated.

Chapter II

DC voltage control strategy for Self exited induction Generator

II. DC voltage control strategy for Self exited induction Generator

II. 1 System description and control scheme

To regulate the voltage of a SEIG with changing load and speed, self-excitation capacitors may be supplemented with an active external source of reactive power. The function of the voltage regulator is to maintain the output voltage of the generator within a given operating range.

The system studied is constituted of a wind turbine, an induction generator, a rectifier/inverter, and loaded with a resistive load by connecting a resistive load, R_{dc}, across the capacitor, C_{dc} as shown Figure 14. The goal of the device is to provide a constant dc voltage to the load connected to the rectifier/inverter even if the speed varies.

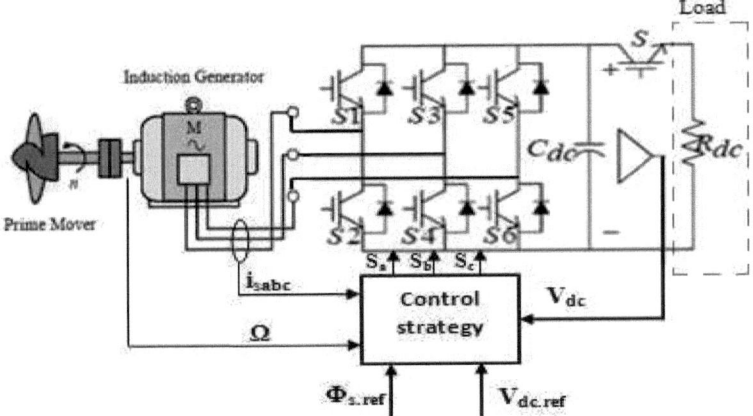

Figure 14: DC voltage control schemes

II.2 PROPOSED CONTROL STRATEGY

The value assumed by the magnetizing inductance depends on the reference magnetic state [12].

$$L_m = \sqrt{\frac{2}{3}} \frac{\Phi_{r \cdot ref}}{|i_m|} = \frac{\sqrt{2}}{3I_s} \Phi_{r \cdot ref} \tag{28}$$

29

On the other hand, the relation between the stator flux and the rotor flux represents a low pass with time constant σT_r :

$$\overline{\Phi}_r = \frac{M}{L_s} \frac{\overline{\Phi}_s}{1 + \sigma T_r s}$$

(29)

From this expression, know the reference stator flux; we can deduce the reference rotor flux. The value assumed by the magnetizing inductance depends on the reference magnetic state. In our case, the stator flux is constant; we can note that L_m is the mutual inductance M introduced in the model of the machine.

II.2.1 Rectifier model

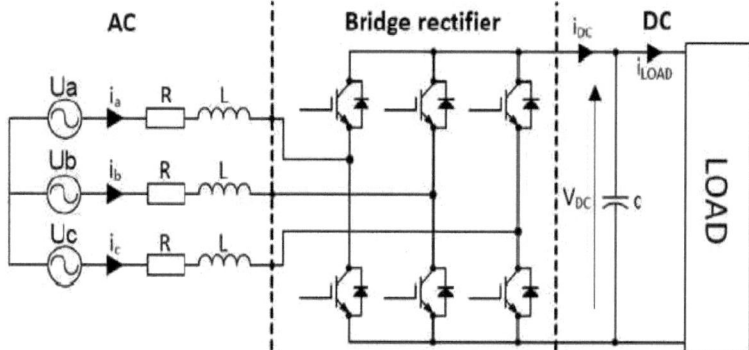

Figure 15: Voltage source AC/DC line-side converter

No neutral connection, we obtain these equations:

$$i_a + i_b + i_c = 0$$

(30)

$$u_{ab} = (S_a - S_b)V_{dc}$$
$$u_{bc} = (S_b - S_c)V_{dc}$$
$$u_{ca} = (S_c - S_a)V_{dc}$$

(31)

With S_i the switching function defined by:

$$S_i = \begin{cases} 0 & \text{Upper switch ON} \\ 1 & \text{Upper switch OFF} \end{cases}$$

With phase i=a, b, c.

$$\begin{aligned} v_{sa} &= m_a . V_{dc} \\ v_{sb} &= m_b . V_{dc} \\ v_{sc} &= m_c . V_{dc} \end{aligned} \tag{32}$$

$$\begin{aligned} m_a &= S_a - \frac{1}{3}(S_a - S_b - S_c) = \frac{2S_a - (S_b + S_c)}{3} \\ m_b &= \frac{2S_b - (S_a + S_c)}{3} \\ m_c &= \frac{2S_c - (S_a + S_b)}{3} \end{aligned} \tag{33}$$

(m_{abc} are 0, ±1/3 or ±2/3)

The rectifier is defined by four equations, one for each phase (voltage) and one for the currents (DC-link), [22]:

$$\begin{bmatrix} v_a \\ v_b \\ v_c \end{bmatrix} = R \begin{bmatrix} i_a \\ i_b \\ i_c \end{bmatrix} + L \frac{d}{dt} \begin{bmatrix} i_a \\ i_b \\ i_c \end{bmatrix} + \begin{bmatrix} v_{sa} \\ v_{sb} \\ v_{sc} \end{bmatrix} \tag{34}$$

$$C_{dc} \frac{dV_{cd}}{dt} = S_a i_a + S_b i_b + S_c i_c - i_{load} \tag{35}$$

Applying the Clarke transformation we can find the voltage equations in αβ-coordinates

$$\begin{bmatrix} u_\alpha \\ u_\beta \end{bmatrix} = R \begin{bmatrix} i_\alpha \\ i_\beta \end{bmatrix} + L \frac{d}{dt} \begin{bmatrix} i_\alpha \\ i_\beta \end{bmatrix} + \begin{bmatrix} u_{s\alpha} \\ u_{s\beta} \end{bmatrix} \tag{36}$$

$$C_{dc}\frac{dV_{cd}}{dt} = \frac{3}{2}(S_\alpha i_\alpha + S_\beta i_\beta) - i_{load} \tag{37}$$

DC load:

The current through the DC load resistance connected across the DC capacitor of the inverter is given by:

$$i_{load} = \frac{V_{dc}}{R_{dc}} \tag{38}$$

II.2.2 Direct Torque and Flux Control (DTFC) Algorithm

This type of control is based on the directly determination of the sequence of control applied to the switches of a tension inverter. This choice is generally based on the use of hysteresis regulators, whose function is to control the state of the system, and to modify the amplitude of the stator flux and the electromagnetic torque.

The stator flux, as given in equation (39), can be approximated as equation (40) over a short time period if the stator resistance is ignored.

$$\overline{\Phi}_s = \overline{\Phi}_{so} + \int_0^t (\overline{V}_s - R_s \overline{I}_s) dt \tag{39}$$

$$\overline{\Phi}_s \approx \overline{\Phi}_{so} + \int_0^t \overline{V}_s dt \tag{40}$$

During one period of sampling T_e, vector tension applied to the machine remains constant, and thus one can write

$$\overline{\Phi}_s(k+1) \approx \overline{\Phi}_s(k) + \overline{V}_s.T_e \tag{41}$$

Or

$$\Delta\overline{\Phi}_s \approx \overline{V}_s.T_e \tag{42}$$

The voltage vector *Vs* is delivered by a three-phase voltage inverter, whose state of the switches are supposed perfect, is represented in theory by three (3) Boolean sizes of control S_j (j=a,b,c):

$$\vec{V}_s = \sqrt{\frac{2}{3}}V_{dc}(S_a + S_b e^{j\frac{2\pi}{3}} + S_c e^{j\frac{4\pi}{3}}) \tag{43}$$

The combinations of the three (3) sizes (S_a, S_b, S_c) make it possible to generate eight (8) positions of the voltage vector Vs whose two (2) positions correspond to the zero vector:

$$(S_a \, S_b \, S_c) = (1 \ 1 \ 1) \quad \text{or} \quad (0 \ 0 \ 0).$$

We use Concordia transformation [23]:

$$\begin{cases} i_{s\alpha} = \sqrt{\dfrac{3}{2}} i_{sa} \\[2mm] i_{s\beta} = \sqrt{\dfrac{1}{2}}(i_{sb} - i_{sc}) \end{cases} \tag{44}$$

$$\begin{cases} V_{s\alpha} = \sqrt{\dfrac{3}{2}} V_{dc}(S_a - \dfrac{1}{2}(S_b - S_c)) \\[2mm] V_{s\beta} = \sqrt{\dfrac{1}{2}} V_{dc}(S_b - S_c) \end{cases} \tag{45}$$

With: S_j (j=a,b,c) are the Boolean sizes of control.

The magnitude of the stator flux is estimated from its components along the axes α and β;

$$\begin{cases} \Phi_{s\alpha} = \displaystyle\int_0^t (V_{s\alpha} - R_s i_{s\alpha})dt \\[2mm] \Phi_{s\beta} = \displaystyle\int_0^t (V_{s\beta} - R_s i_{s\beta})dt \end{cases} \tag{46}$$

$$\Phi_s = \sqrt{\Phi_{s\alpha}^2 + \Phi_{s\beta}^2} \tag{47}$$

Figure 16 gives the strategy of control proposed:

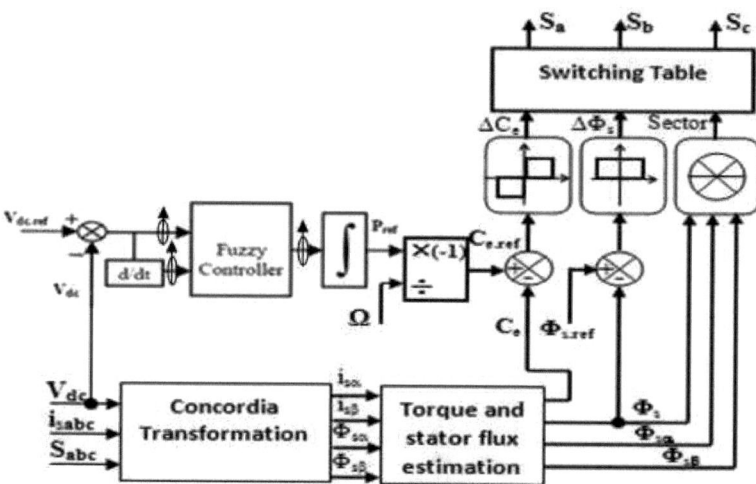

Figure 16: Control structure proposed

Therefore to increase the stator flux, we can apply a vector of tension that is co-linear in its direction and vice-versa.

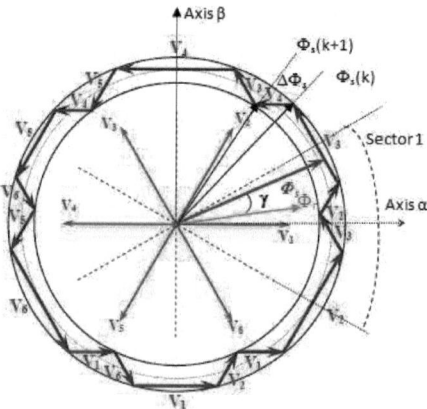

Figure 17: Definition of stator flux increment and spatial positions of the voltage vectors keeping the flux inside the strip of hysteresis.

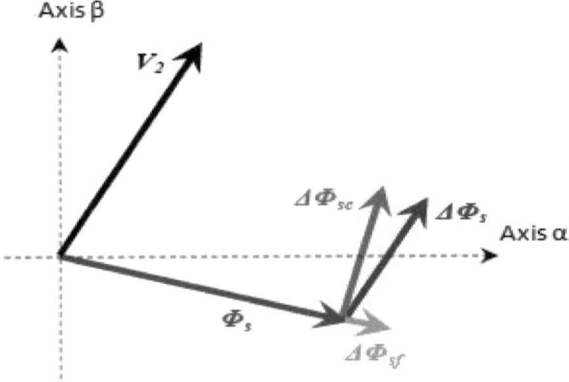

Figure 18: Components of the error of flux at the time of the application of the vector V_2 voltage

If the error of flux is projected on the direction of stator flux and on a perpendicular direction (Fig.18), one puts in evidence the components acting on the torque and on the flux.

In the Figure 18, the component $\Delta\Phi_{sc}$ gives the electromagnetic Torque of the Induction motor while the component $\Delta\Phi_{sf}$ modifies the magnitude of stator flux.

The electromagnetic torque can be estimated starting from the estimated sizes of flux ($\Phi_{s\alpha}$ and $\Phi_{s\beta}$) and the calculated sizes of current ($i_{s\alpha}$ and $i_{s\beta}$) (Fig.16).

$$C_{em} = p(\Phi_{s\alpha}i_{s\beta} - \Phi_{s\beta}i_{s\alpha})$$

(48)

When flux is in sector S_i, the vectors V_{i+1} or V_{i-1} are selected to increase the amplitude of flux, and V_{i+2} or V_{i-2} to decrease it. What shows that the choice of the vector tension depends on the sign of the error of flux, independently of its amplitude. This explains why the exit of the corrector of flux can be a Boolean variable. One adds a bond of hysteresis around zero to avoid useless commutations when the error of flux is small. Indeed, with this type of corrector in spite of his simplicity, one can easily control and maintain the end of the vector flux, in a circular ring. The switching table proposed by Takahashi [21], as given by Table 2.

$\Delta\phi_s$	ΔC_e	S_1	S_2	S_3	S_4	S_5	S_6
	1	110	010	011	001	101	100
1	0	000	000	000	000	000	000
	-1	101	100	110	010	011	001
	1	010	011	001	101	100	110
0	0	000	000	000	000	000	000
	-1	001	101	100	110	010	011

Table2: Switching table

II.2.3 Voltage Controller Design

The main task of the induction generator is to regulate the dc-link voltage. The voltage control is carried out through a voltage control loop using a Fuzzy PI controller.

From the value of the desired voltage, it is possible to express the reference power:

$$V_{dc.ref} . i_{dc} = P_{ref}$$

(49)

Where i_{dc} is the rectifier output current.

Neglecting the various losses, we obtain directly the expression of electromagnetic torque by:

$$C_{ref} = \frac{P_{ref}}{\Omega}$$

(50)

Control voltage V_{dc} can be done via the electromagnetic torque control, which amounts to the same approach as that used in the case of a conventional motor control.

II.2.4 Fuzzy Logic Controller

Fuzzy Logic is based on the theory of fuzzy sets developed by Zadeh [15]. This is an extension of the classical theory for the incorporation of fuzzy set. Fuzzy control was found particularly useful to solve nonlinear control problems or when the plant model is unknown or difficult to build. The design of the speed controller is based on the experience and intuition of human plant operator

36

without knowing the mathematical model of the Motor. The fuzzy controller generates the incremental control signal error from error E and dE.

The proposed Fuzzy controller has two inputs and one output as described in figure 19.

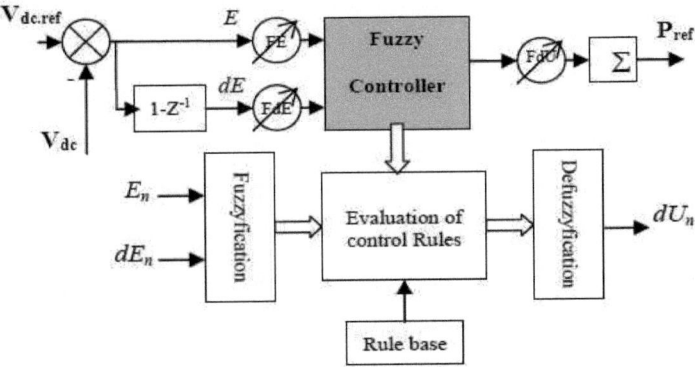

Figure 19: Structure of the Fuzzy controller

Where **E** is the error, expressed by:

$$E(k) = V_{dc.ref}(k) - V_{dc}(k-1)$$
(51)

dE is derived from the error approximated by:

$$dE(k) = \frac{E(k) - E(k-1)}{T_e}$$
(52)

With Te is the sampling period.

The output of the regulator is given by:

$$P_{ref}(k) = P_{ref}(k-1) - dU(k)$$
(53)

FE, FdE, FdU are gains called "scale factor". They can change the sensitivity of the controller without changing its structure.

The fuzzy controller is composed of three blocks: Fuzzification, rule bases, and Defuzzification. Figure 20 show the function of membership of each input signals (E, dE). The fuzzy subsets are as follows:

NB (Negative Big), Nm (Negative Medium), NS (Negative Small), Z (Zero), PS (Positive Small), PM (Positive Medium), PB (Positive Big).

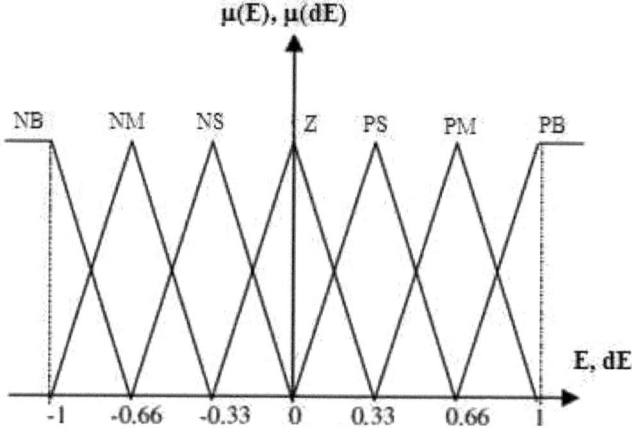

Figure 20: Membership function

There are 7 fuzzy subsets for each variable, which gives 7 * 7 = 49 possible rules, where a typical rule is:

"If **E** is **PS** and **dE** is **PM** Then **dU** is **PB** ".

dEn \ En	NB	NM	NS	Z	PS	PM	PB
PB	Z	PS	PM	PB	PB	PB	PB
PM	NS	Z	PS	PM	PB	PB	PB
PS	NM	NS	Z	PS	PM	PB	PB
Z	NB	NM	NS	Z	PS	PM	PB
NS	NB	NB	NM	NS	Z	PS	PM
NM	NB	NB	NB	NM	NS	Z	PS
NB	NB	NB	NB	NB	NM	NS	Z

Table 3: rules base

Defuzzification is done by centroid method based on the inference method Takagi-Sugeno-Kang.

II.3 Simulation results

The induction machine used as the SEIG in this investigation is a three-phase squirrel cage induction generator with specification:

3Kw, $\cos\varphi = 0.8$, 4 poles, 220V (rms). $C_{dc} = 1\mu$Farad.

See appendix for other parameters.

References controlled systems are:

$\Phi_{s.ref} = 0.85$Wb

$V_{dc.ref} = 400$V

$\Omega_{ref} = 160$ rad/s

The DC voltage regulation is obtained using the proposed algorithm controller in spite of the presence of disturbances such as step changing of the resistive load and the mechanical speed (when the SEIG is driven by a wind turbine).

II.3.1 functioning with changing of the mechanical speed

Figure 21 to Figure 26 shows a no-load operation flowed-up by a step changing of the mechanical speed with an increase at $t=1.5s$ and a decrease at $t=3s$, when the DC-bus voltage is set to 400V.

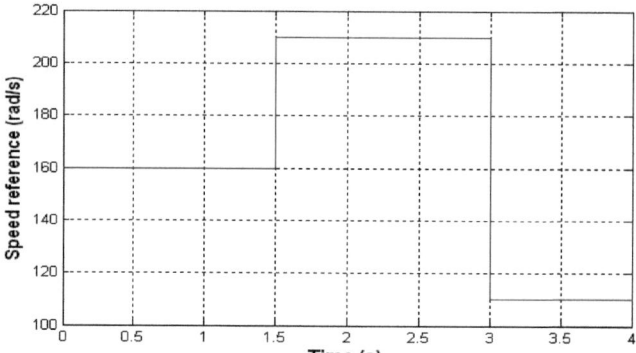

Figure 21: Mechanical speed variation

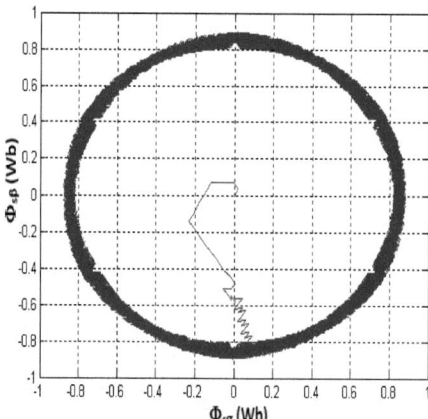

Figure 22: stator flux trajectory

Figure 22 show the plan (α, β), the stator flux is driven in a circle module 0.85wb.

In figure 23 and figure 24, applying guidelines as a DC voltage one of 400V and stator flux of 0.85Wb. Clearly, the system follows these guidelines. Fluctuations current and flux are due to the comparator hysteresis.

The shape of the current is sinusoidal representative fluctuations of the order of 10%, around the average value instantaneous, also due to the comparator hysteresis.

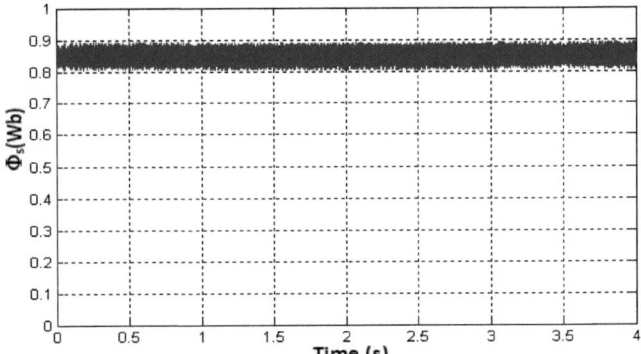

Figure 23 : Magnitude of the stator flux

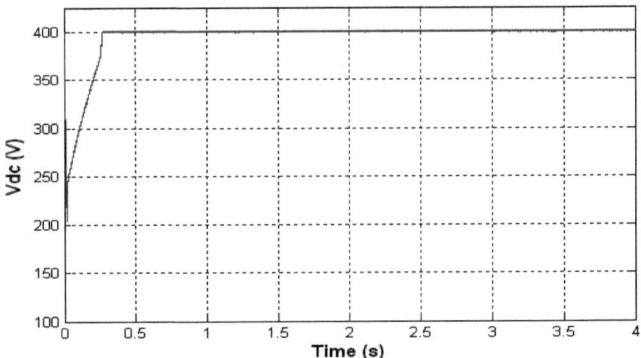

Figure 24 : DC voltage

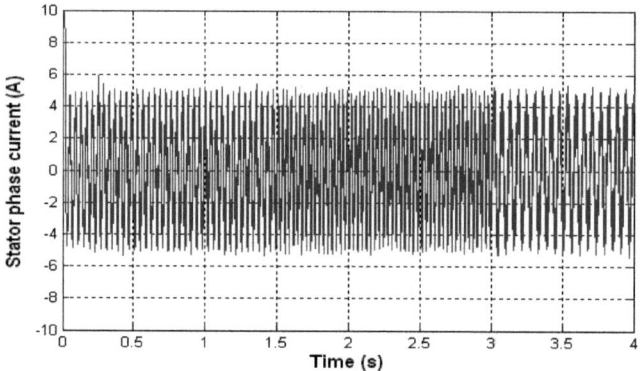

Figure 25 : Stator phase current

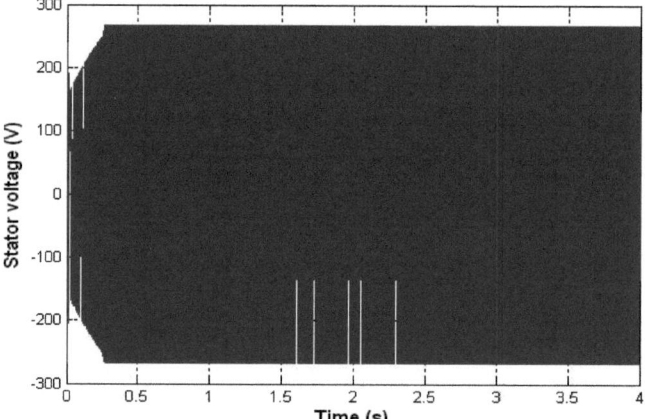

Figure 26: Generated voltage at the terminal of induction generator

II.3.2 Load application and removal on DC side

To investigate the response of the system with sudden application and removal of load on the DC side, the generator is initially excited at no-load and suddenly a DC load of (R_{dc} = 150Ω) is applied at t = 1s, and this load is removed at t= 1.5s, as shown in Figure 27.

At the time of application of load, the DC voltage decrease but quickly return to reference values. When the load is removed, the phase voltage of the generator and the DC voltage increase due to mismatch in active power produced by the SEIG, which is more than the power consumed by the load. The duty cycle of the inverter is adjusted by the control action and the voltages are maintained at their reference values.

Figure 28 and Figure 29 shows the stator phase current and the buildup of generated voltage at the terminals of the induction generator in the conditions of Figure 27. In these results, we note that the voltage comes to the reference value quickly. The phase current of the SEIG increases at the time of application of load.

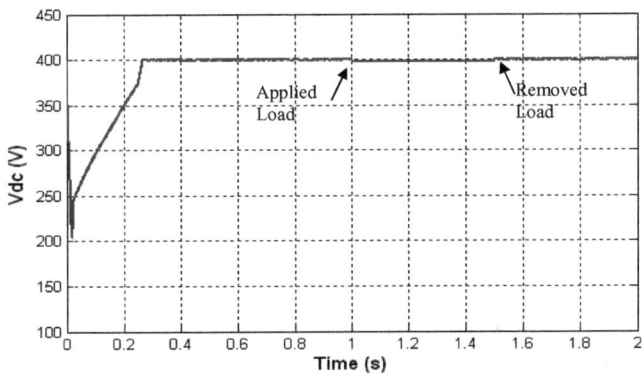

Figure 27 : DC voltage

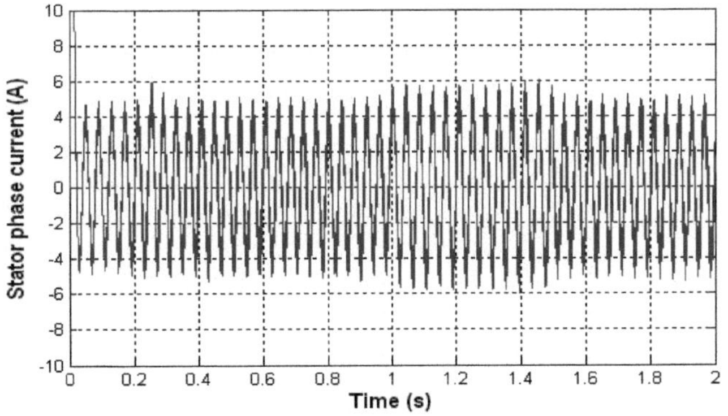

Figure 28 : Stator phase current

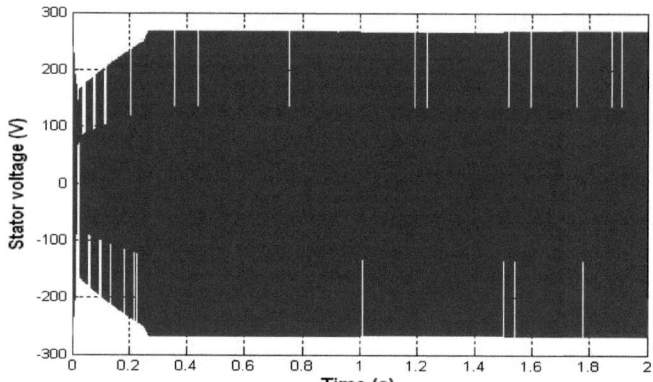

Figure 29: Generated voltage at the terminal of induction generator

II.2.4 Summary

In this second chapter one proposed the DC voltage regulation of an induction generator by a fuzzy Logic controller ordered by direct torque and flux control strategy. The structures studied have been implemented using MATLAB/SIMULINK. With mechanical speed variation and sudden application and removal of load, the fuzzy controller provides good dynamic performances.

The excellent results obtained show the effectiveness of this technique in the regulation DC voltage of the self-excited induction generator used in wind turbine.

CONCLUSION

The main advantages of renewable are available, clean, low cost and continuous energy. The reasons for choosing induction generator in wind energy system are that its very reliable tends to be comparatively inexpensive, light weight, and low maintenance. The generator also has some mechanical properties which are useful for wind turbines.

This book examines the phenomenon of self-excitation in an induction generator which is of practical interest. Therefore the advanced knowledge of the minimum excitation capacitor value is required .To find these capacitor value two non-linear equations have to be solved. The formula to calculate the minimum capacitance required for self-excited induction generator is simple and it doesn't need numerical iteration. For this reason, this formula helps to determine the minimum capacitance required for self-excited induction generator on line. The formula gives typical results as the results obtained from iterative technique without any iteration or divergence problem. For stable operation C must be chosen to be slightly greater than Cmin.

The modeling, analysis and dynamic performance of an isolated three-phase induction generator excited by three AC capacitors connected at the stator terminals is presented in Chapter I. The use of the variation in magnetising inductance with voltage leads to an accurate prediction of whether or not self-excitation will occur in a SEIG for various capacitance values and speeds in both the loaded and unloaded cases. The characteristics of magnetising inductance, L_{mg}, with respect to the rms induced stator voltage or magnetising current determines the regions of stable operation as well as the minimum generated.

When an induction machine operates as a motor the speed of the rotating air gap magnetic field is totally dependent on the excitation frequency. However, in the SEIG it is shown that the frequency of the generated voltage depends on the speed of the prime mover as well as the condition of the load. Keeping the speed of the prime mover constant with increased load causes the magnitude of generated voltage and frequency of an isolated SEIG to decrease.

Chapter II presents a simple algorithm based on the direct torque and flux control (DTFC) for Self-excited induction generators (SEIG) with a Fuzzy logic controller scheme. This strategy has been presented to maintain the terminal voltage of the generator and the DC bus voltage constant for variable rotor speed and load.

The Fuzzy control was found particularly useful to solve nonlinear control problems or when the plant model is unknown or difficult to build. In this chapter, it will be shown that these techniques can also be useful in applications where classical control performs well. Fuzzy Logic allows a simpler and more robust control solution whose performance can only be matched by a classical controller with adaptive characteristics, much more difficult to implement.

The proposed control system provides excellent DC voltage control with fast transient response and no steady state error.

APPENDIX

The machine used in this work is the induction machine characterized by nominal values:

3 kW, 1400 rpm, 220/380 V, 12.5/7.2 A, 3 phases, 50 Hz.

1.5KW, 1440 rpm, 220/380V, 8/5.6 A, 3 phases, 50 Hz

The parameters of the used induction machines are summarized in table 4 and table 5, they were obtained by using the laboratory testing as described in [23].

Rated power	3 KW
Voltage	380V Y
Frequency	50 Hz
Pair pole	2
Rated speed	1400 rpm
Stator resistance	2.2 Ω
Rotor resistance	2.68 Ω
Inductance stator	229 mH
Inductance rotor	229 mH
Mutual inductance	217 mH
Moment of Inertia	0.046 kg.m^2

Table 4: Induction generator parameters

Rated power	1.5 KW
Voltage	380V Y
Frequency	50 Hz
Pair pole	2
Rated speed	1440 rpm
Stator resistance	4.85 Ω
Rotor resistance	3.805 Ω
Inductance stator	274mH
Inductance rotor	274 mH
Mutual inductance	258 mH
Moment of Inertia	0.031 kg.m^2

Table 5: Induction motor parameters

REFERENCES

[1] C. Chakraborty, S.N. Bhadra, A.K. Chattopadhyay,"Excitation requirements for standalone three phase induction generator", IEEE Transactions on Energy Conversion, Vol.13, No. 4, 1998, pp 358-365.

[2] G. Rains and 0 P. Malik, "Wind energy conversion using a self-excited induction generator," IEEE Transactions on Power Apparatus and Systems. vol.102, no. 12, 1983, pp. 3933-3936.

[3] A.H. AI-Bahrani "Analysis of self-excited induction generators under unbalanced conditions."Electric Machine and Power Systems. vol 24.1996,pp 117-129.

[4] M. Radic, Z. Stajic, D. Arnautovic," critical speed-capacitance requirements for self-excited induction generator", Automatic Control and Robotics Vol. 8, No 1, 2009, pp. 165 – 172.

[5] A. Kishore, G. Satish Kumar," dynamic modeling and analysis of three phase self-excited induction generator using generalized state-space approach", International Symposium on Power Electronics, Electrical Drives, Automation and Motion, SPEEDAM 2006, pp 52-59.

[6] Dawit S Eyaum Colin Grantham, and Muhammed Fazlur Rahman,"The dynamic characteristics of an isolated self-excited induction generator driven by a wind turbine", IEEE, Transaction on Industry applications, Vol. 39.No.4 , July/August 2003 pp 936 - 944.

[7] R. Kumar T., V. Agarwal, P.S.V. Nataraj," A Reliable and Accurate Calculation of Excitation Capacitance Value for an Induction Generator Based on Interval Computation Technique", International Journal of Automation and Computing 8(4), November 2011, pp 429-436.

[8] Li Wang, Chaing-Huei Lee , "A novel analysis on the performance of an Isolated self excited induction generators," IEEE Trans. on Energy conversion June 1993, vol.12, No.2.

[9] Sridhar, L., B. Singh, C. S. Jha, and B. P. Singh, "Analysis of a self excited induction generator feeding induction motor load," IEEE Transaction on Energy Conversion, Vol. 9, No. 2, June 1994.

[10] S. Boora," On-Set Theory of Self-Excitation in Induction Generator", International Journal of Recent Trends in Engineering, Vol 2, No. 5, November 2009,pp 325-330.

[11] M. Ouali, M. Ben ali Kamoun, M. Chaabene," Investigation on the Excitation Capacitor for a Wind Pumping Plant Using Induction Generator", Smart Grid and Renewable Energy, 2011, 2, pp 116-125.

[12] R. O. C. Lyra, S. R. Silva and P. C. Cortizo, "Direct and indirect flux control of an isolated induction generator", Proceedings of IEEE International Conference on Power Electronics and Drive Systems, Vol. 1, pp. 140-145, 1995.

[13] D. Seyoum, M. F. Rahman and C. Grantham, "Terminal voltage control of a wind turbine driven isolated induction generator using stator oriented field control", Proceedings of IEEE Power Electronics Conference and Exposition- APEC, Vol. 2, pp. 846-852, 2003.

[14] R. Leidhold, G. Garcia and M I. Valla, "Field-oriented controlled induction generator with loss minimization", IEEE Transactions on Industrial Electronics, Vol. 49, No. 1, pp. 147-156, February 2002.

[15] L. A. Zadeh, "Fuzzy Setes" Information and Control, vol. 8, pp.338-353, 1965.

[16] A. Abbou , Y. Sayouti, H. Mahmoudi, and M. Akherraz ," dSPACE Direct Torque Control Implementation for Induction Motor Drive", Proceedings in the 18th IEEE Mediterranean Conference on control and Automation, Congress Palace, Marrakech, Morocco, ,pp. 1121-1126, 2010.

[17] BLASKO V., KAURA V. "A New Mathematical Model and Control of a Three-Phase AC–DC Voltage Source Converter". IEEE Transactions on Power Electronics, 1997. Vol.12 , issue: 1. Pages: 116 –123, 1997.

[18] D. Rekioua, T. Rekioua, S. Alloune, "Switching Strategies in Direct Torque Control of Induction Machine: Modelling and simulation", International Conference Modelling And Simulation (MS2004)-Lyon France, 4-7 Juillet 2004, pp : 3-18-3-21,2004.

[19] K. Idjdarene, D. Rekioua, T. Rekioua and A. Tounsi, "Control strategies for an autonomous induction generator taking the saturation effect into account," European Conference on Power Electronics and Applications, pp. 1-10, 2-5 Sept. 2007.

[20] T. Ahmed, K. Nishida and M. Nakaoka, "A novel standalone induction generator system for AC and DC poer applications", IEEE Transactions on Industry Applications, ol. 43, No. 6, pp. 1465-1474, November/December 2007.

[21] I Takahashi, Direct Y.Ohmori, "High-performance Torque Control of year Induction Motor", IEEE Transactions one Industry Applications, vol. 25, pp. 257-264, March/April 1989.

[22] J. Holtz, Sensorless Control of induction motor drives, Proc.IEEE, vol. 90, n°8, pp.1359-1394, 2002.

[23] A. Abbou and H. Mahmoudi ,"Real Time Implementation of a Sensorless Speed Control of Induction Motor using DTFC Strategy", Proceedings of the International Conference on Multimedia Computing and Systems ICMCS'09, pp.327-333, Ouarzazate, Morocco, April 2-4, 2009.

[24] Kamalzadeh, Abdulrahim; Radan, Ahmad," Direct Power Control of Doubly-Fed Induction Generators Using an Analytical Optimized Switching Table",IREE, vol.5, Issue 1, pp.70-82,2010.

[25] Mihoub, Y.; Toumi, D.; Mazari, B.; Hassaine, S.," Design and Implementation of an Adaptive PI Fuzzy Controller to Improve the Speed Control of Induction Motor",IREE, vol.5, Issue 2, pp.481-490,2010.

[26] Kessal Abdelhalim, Rahmani Lazhar, Jean-Paul Gaubert, Mostefai Mohammed,' Hysteresis-band Current Control of PFC with Constant Switching Frequency', IREE, vol.6, Issue 1, pp.179-185,2011.

[27] N. Gupta, S. P. Singh, S. P. Dubey, D. K. Palwalia," Fuzzy Logic Controlled Three-phase Three-wired Shunt Active Power Filter for Power Quality Improvement", IREE, vol.6, Issue 3, pp.1118-1129,2011.

I want morebooks!

Buy your books fast and straightforward online - at one of the world's fastest growing online book stores! Environmentally sound due to Print-on-Demand technologies.

Buy your books online at

www.get-morebooks.com

Kaufen Sie Ihre Bücher schnell und unkompliziert online – auf einer der am schnellsten wachsenden Buchhandelsplattformen weltweit!
Dank Print-On-Demand umwelt- und ressourcenschonend produziert.

Bücher schneller online kaufen

www.morebooks.de

OmniScriptum Marketing DEU GmbH
Heinrich-Böcking-Str. 6-8
D - 66121 Saarbrücken
Telefax: +49 681 93 81 567-9

info@omniscriptum.com
www.omniscriptum.com

MIX
Papier aus verantwortungsvollen Quellen
Paper from responsible sources
FSC® C105338

Printed by Books on Demand GmbH, Norderstedt / Germany